SNAKES ALIVE
Coral Snakes

by Colleen Sexton

BELLWETHER MEDIA • MINNEAPOLIS, MN

Note to Librarians, Teachers, and Parents:

Blastoff! Readers are carefully developed by literacy experts and combine standards-based content with developmentally appropriate text.

Level 1 provides the most support through repetition of high-frequency words, light text, predictable sentence patterns, and strong visual support.

Level 2 offers early readers a bit more challenge through varied simple sentences, increased text load, and less repetition of high-frequency words.

Level 3 advances early-fluent readers toward fluency through increased text and concept load, less reliance on visuals, longer sentences, and more literary language.

Level 4 builds reading stamina by providing more text per page, increased use of punctuation, greater variation in sentence patterns, and increasingly challenging vocabulary.

Level 5 encourages children to move from "learning to read" to "reading to learn" by providing even more text, varied writing styles, and less familiar topics.

Whichever book is right for your reader, Blastoff! Readers are the perfect books to build confidence and encourage a love of reading that will last a lifetime!

This edition first published in 2011 by Bellwether Media, Inc.

No part of this publication may be reproduced in whole or in part without written permission of the publisher. For information regarding permission, write to Bellwether Media, Inc., Attention: Permissions Department, 5357 Penn Avenue South, Minneapolis, MN 55419.

Library of Congress Cataloging-in-Publication Data

Sexton, Colleen.
 Coral snakes / by Colleen Sexton.
  p. cm. – (Blastoff! readers. Snakes alive!)
 Summary: "Simple text and full-color photography introduce beginning readers to coral snakes. Developed by literacy experts for students in kindergarten through third grade"–Provided by publisher.
 Includes bibliographical references and index.
 ISBN 978-1-60014-540-7 (paperback : alk. paper)
 1. Coral snakes–Juvenile literature. I. Title.
 QL666.O64S49 2010
 597.96'44-dc22

2009037592

Text copyright © 2011 by Bellwether Media, Inc. BLASTOFF! READERS and associated logos are trademarks and/or registered trademarks of Bellwether Media, Inc.

Printed in the United States of America, North Mankato, MN.

Contents

How Coral Snakes Look	4
Escaping Predators	9
Where Coral Snakes Live	12
Hunting and Eating	14
Glossary	22
To Learn More	23
Index	24

Coral snakes are small, colorful snakes. Most grow 18 to 40 inches (46 to 102 centimeters) long. Some stretch as long as 4 feet (1.2 meters).

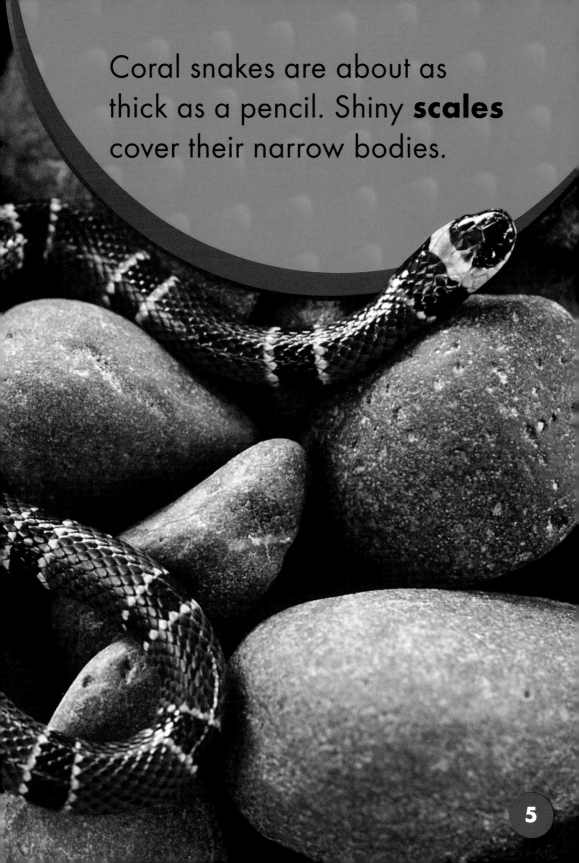

Coral snakes are about as thick as a pencil. Shiny **scales** cover their narrow bodies.

Coral snakes have bright colors. They have red and black bands. They also have either yellow or white bands.

The bright colors warn other animals that coral snakes are **poisonous**.

The colors do not scare away coyotes, foxes, large snakes, and other **predators**. These animals will eat coral snakes if they can catch them.

Coral snakes stay safe from predators in many ways. They hide under leaves or rocks. They escape down holes made by other animals.

Coral snakes also **coil** their bodies. Then they hide their heads and stick up their tails.

A predator grabs the tail. The coral snake then lifts its head and surprises the predator!

= areas where coral snakes live

Coral snakes live in warm parts of North America, Central America, and South America.

Coral snakes can be found in deserts, grasslands, forests, and mountains.

Coral snakes hunt animals that share their **habitat**. They hunt smaller snakes, lizards, and other **prey**.

Coral snakes hunt at night and early in the morning. They stick out their forked tongues to pick up the scent of prey.

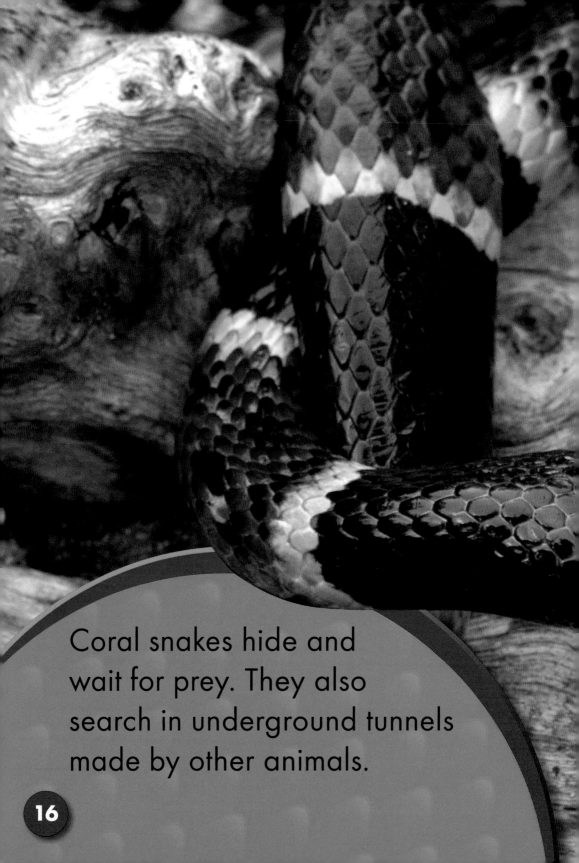

Coral snakes hide and wait for prey. They also search in underground tunnels made by other animals.

Sometimes coral snakes drop their tails into holes to scare out small animals.

Coral snakes **strike** at their prey. They bite other snakes behind the head so that they cannot bite back.

Coral snakes bite with their sharp, curved **fangs**. A poison called **venom** flows through the fangs and into the bite.

The venom makes an animal unable to move or breathe. It soon dies.

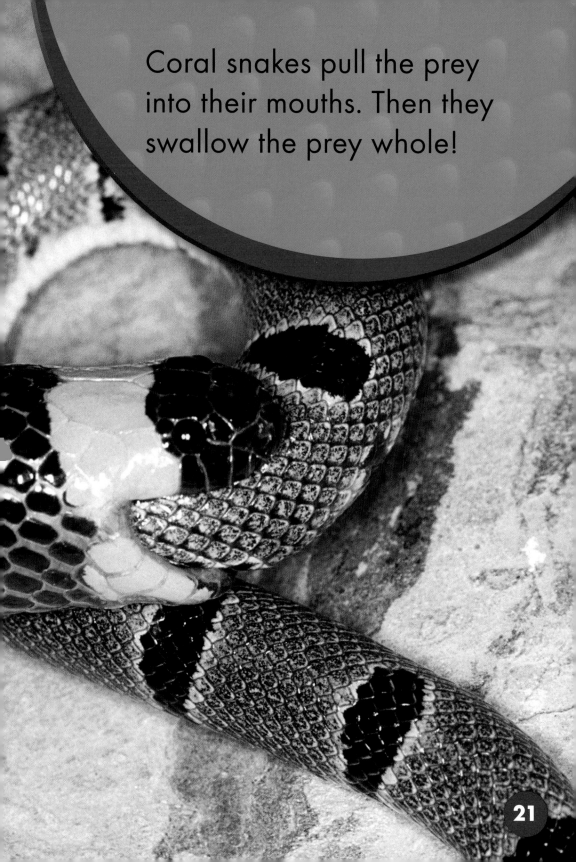

Coral snakes pull the prey into their mouths. Then they swallow the prey whole!

Glossary

coil—to wind into loops

fangs—sharp, curved teeth; coral snakes have hollow fangs through which venom can move into a bite.

habitat—the natural surroundings in which an animal lives

poisonous—able to kill or harm with a poison; the venom that a coral snake makes is a poison.

predator—an animal that hunts other animals for food

prey—an animal hunted by another animal for food

scales—small plates of skin that cover and protect a snake's body

strike—to quickly throw the head and front part of the body at a predator or prey

venom—a poison that some snakes make; coral snake venom is deadly.

To Learn More

AT THE LIBRARY

Gibbons, Gail. *Snakes*. New York, N.Y.: Holiday House, 2007.

Gunzi, Christiane. *The Best Book of Snakes*. New York, N.Y.: Kingfisher, 2003.

Patent, Dorothy Hinshaw. *Slinky, Scaly, Slithery Snakes*. New York, N.Y.: Walker & Co., 2000.

ON THE WEB

Learning more about coral snakes is as easy as 1, 2, 3.

1. Go to www.factsurfer.com.

2. Enter "coral snakes" into the search box.

3. Click the "Surf" button and you will see a list of related Web sites.

With factsurfer.com, finding more information is just a click away.

Index

bands, 6
Central America, 12
coiling, 10
colors, 6, 7, 8
deserts, 13
fangs, 19
forests, 13
forked tongues, 15
grasslands, 13
habitat, 14
hunting, 14, 15
length, 4
mountains, 13
North America, 12
poison, 7, 19
predators, 8, 9, 11
prey, 14, 15, 16, 18, 21

scales, 5
South America, 12
striking, 18
swallowing, 21
tail, 10, 11, 17
venom, 19, 20

The images in this book are reproduced through the courtesy of: Michael_Patricia Fogden/Minden Pictures, front cover, pp. 6-7; Lubeck, Robert/Animals Animals – Earth Scenes, pp. 4-5; Florida Images/Alamy, p. 6 (small); Jon Eppard, p. 8 (small); David M. Dennis, pp. 8-9; Juan Martinez, p. 10; John Cancalosi, p. 11; Minden Pictures, pp. 12-13, 18-19; Kent, Breck P./Animals Animals – Earth Scenes, p. 13 (small); Steve Bower, p. 14 (small); Larry Ditto/bciusa.com, pp. 14-15; James H. Carmichael/bciusa.com, pp. 16-17; Mark Moffett/Minden Pictures, p. 19 (small); Freed, Paul/Animals Animals – Earth Scenes, pp. 20-21.